The Water Cycle with Olly and Carl

Sonja McGiboney

ISBN: 978-1-964345-04-8
Library of Congress Control Number: 2024914732
Published by Sonja McGiboney
Paperback

The Water Cycle with Olly and Carl

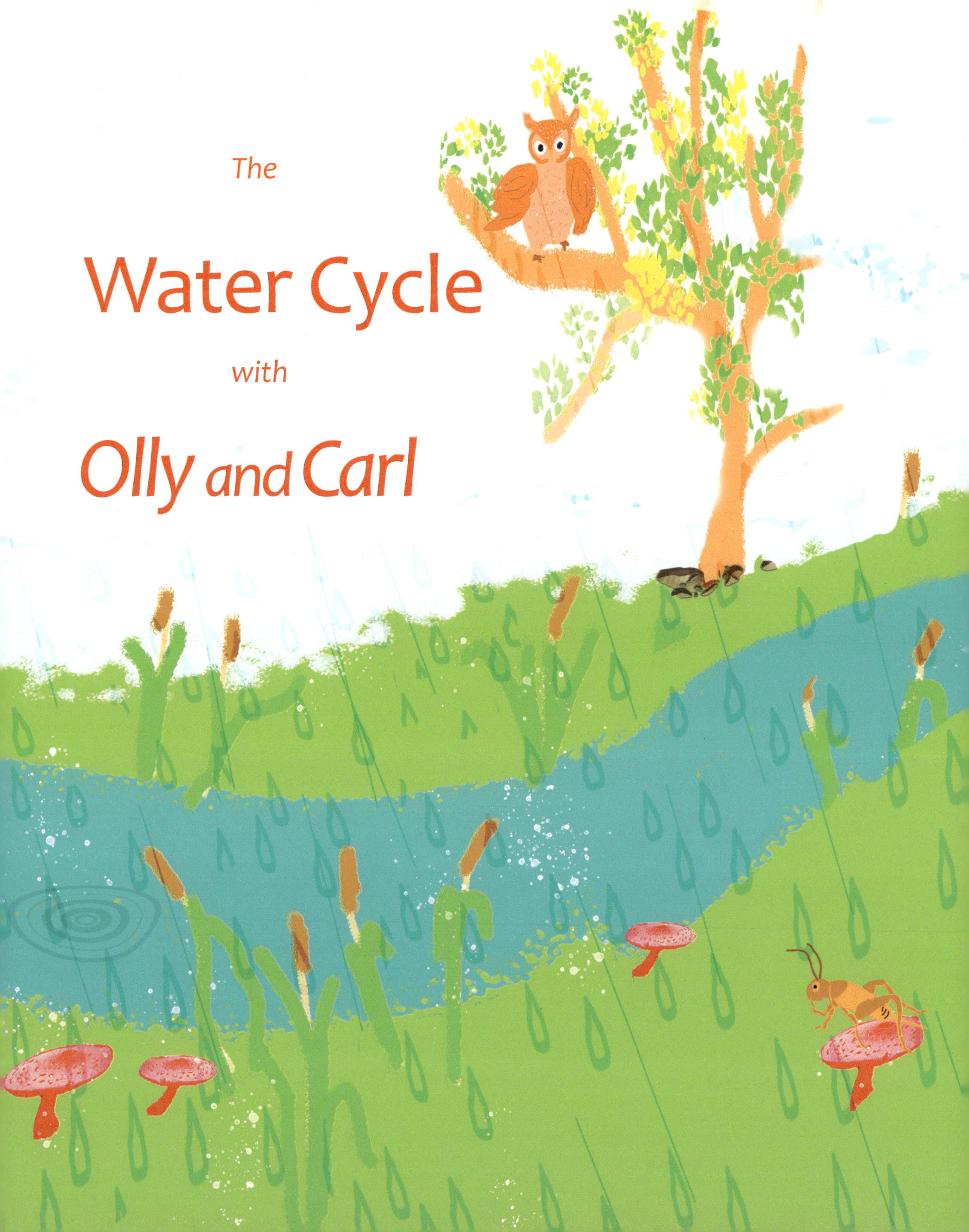

Olly landed on a tree branch and closed his eyes. The rain sang a soft lullaby. "Pitter-patter, pitter-patter."

"Creek-reek, creek-reek, boo-hoo, boo-hoo!" A loud cry came from below.

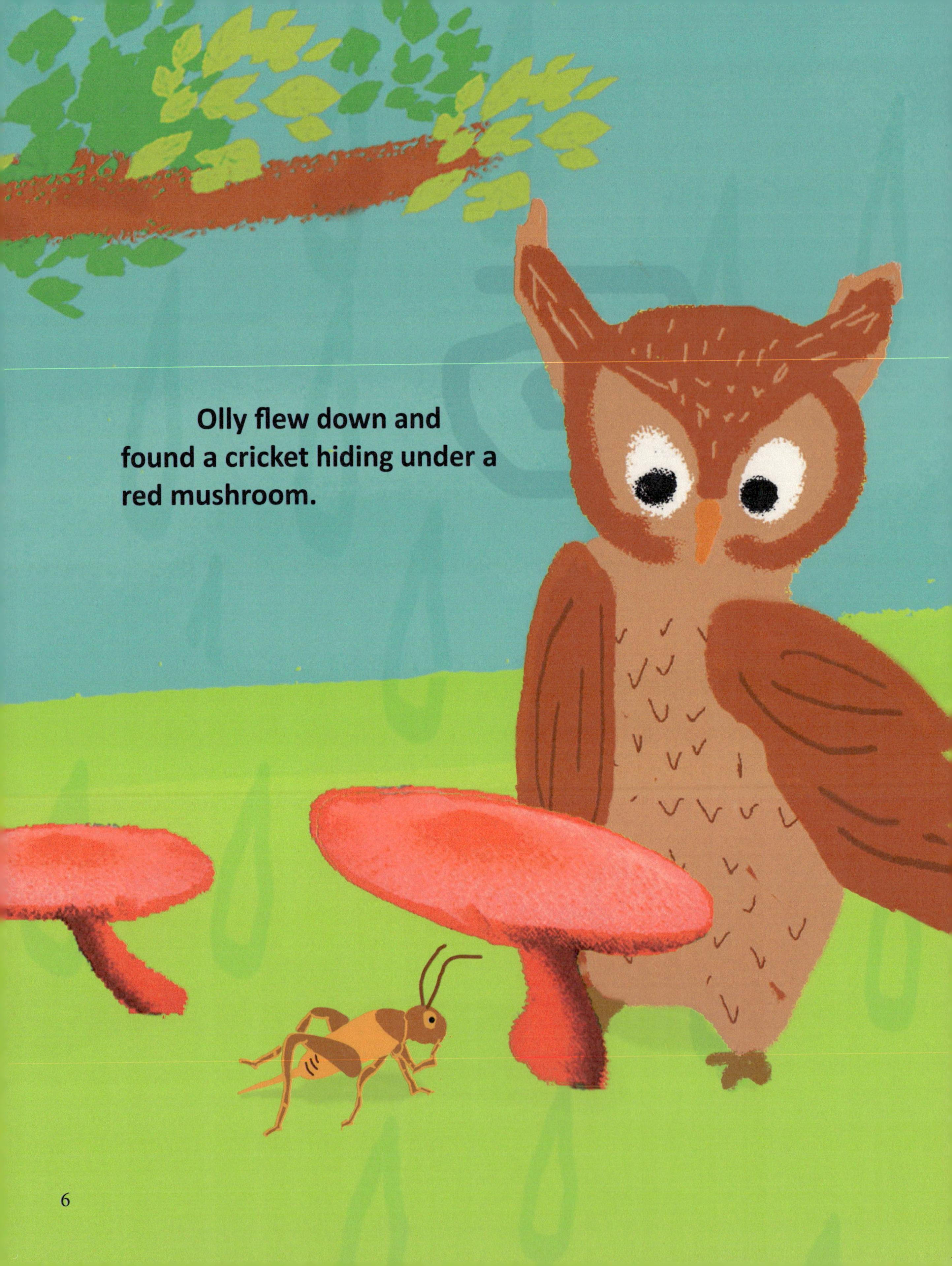

Olly flew down and found a cricket hiding under a red mushroom.

"Who are you, and why are you crying?"

"I'm Carl, and I'm scared. If I get wet, I will melt. *Sniff!* Please make it stop!"

"Hello, Carl. My name is Olly. Why do you think you will melt in the rain?"

Carl remembered his bad dream about melting in the water. He told Olly all about it.

Olly felt sad. “Oh, Carl, I’m sorry you had a bad dream, but there is nothing to be afraid of.”

Olly spread his wing like an umbrella and turned to face the river. "Are you afraid of the river?"

"No. It's where I wash and drink."

"Did you know that rain comes from the river?"

Carl's eyes widened in surprise. "It does?"

"Yes, Carl. The heat from the sun causes the water in the river to get hot and it becomes air."

"Is that magic?" asked Carl.

Olly laughed. “No, it’s not magic. It’s called EVAPORATION.”

“EVAPRASHUN?” echoed Carl.

“No, listen. EEEVAAPOOORRAATION!”

Carl practiced by repeating the word, “Evaporation, evaporation, evaporation.” Then he said, “How come I can’t see it!”

“Because evaporated water is a gas called water vapor. We can’t see it, but we can feel it.”

"What about the puddles? Do they evaporate, too?"

"Yes, but some of the water goes into the ground. Plants and trees use their roots to get water that goes into their stems, leaves, and flowers. If there is too much water, it goes back into the air. That's called transpiration."

Carl asked, "Where does the evaporation and tran-spay-shun go?"

"You mean evaporation and transpiration?"

"Yeah, that."

Olly pointed to the clouds. “The clouds are like big balls of evaporated water. Way up in the sky, the air is colder. When there is too much water for the air to hold, the water vapor turns into water droplets. It’s called CONDENSATION!”

Carl tried to say it, “CONE-DANCING!”

“No,” Olly said, “try it again. CON-DEN-SA-TION!”

Carl said the words. "It sounds like a choo-choo train. Condensation, transpiration, condensation, transpiration, condensation, transpiration." He smiled.

“There’s even another word to learn. The word is PRECIPITATION,” said Olly.

“What is that?”

“PRECIPITATION is the different ways water falls from the sky. Right now it is rain, but it could be snow, sleet or hail. ”

Carl hopped on top of the red mushroom. "Are you sure it isn't magic?"

Olly laughed. "Yes, I'm sure it is not magic. It is the water cycle. Water goes around and around, through all the stages:

EVAPORATION,
TRANSPIRATION,
CONDENSATION,
and PRECIPITATION!"

Carl danced in circles on the mushroom, repeating all the words. "Evaporation, transpiration, condensation, precipitation."

"Carl?" Olly raised his wing.

"Yes, Olly?"

"Are you going to be afraid of the rain anymore?"

"No," said Carl, "I like the rain. I like the water cycle!"

Olly flew up to the tree and landed on the branch. He closed his eyes, thinking, "Now I can finally get some sleep."

But then he heard a different noise.

Carl, no longer afraid of the rain, sang out his happiness. "La, la, la, la, it's only water. La, la, la, la!"

Poor Olly. Will he ever get to sleep?

Glossary

Here are more words that are associated with the water cycle and their definitions.

The Water Cycle: Sometimes called the **Hydrologic Cycle**, it is the continuous movement of water through and around the earth.
Water Vapor: The gas formed from heated or boiled water.
Evaporation: The sun heats liquid water and it turns into water vapor.
Condensation: Water vapor cools and forms liquid water.
Precipitation: Water returns to the earth's surface as rain, sleet, snow, or hail.

Transpiration: The sun pulls water from the plants through leaf openings called stomata.

Runoff: Water, pulled by gravity, gets pulled over the ground.
Infiltration: When water seeps into the ground.
Storage: Water is stored three basic locations: in the atmosphere; on the surface of the earth as oceans, lakes, and rivers; and in the ground.

You can find out more information about the water cycle in encyclopedias and dictionaries, on websites, or from your teachers.

Here is where I found my information.

Corps of Engineers, Portland District, article accessed on 6-16-2024

{https://www.nwrfc.noaa.gov/info/water_cycle/hydrology.html#:~:text=It%20can%20be%20studied%20by,transpiration%2C%20runoff%2C%20and%20storage.&text=Evaporation%20occurs%20when%20the%20physical,state%20to%20a%20gaseous%20state.}

National Geographic Kids, article accessed on 6-16-2024

{https://www.natgeokids.com/uk/discover/science/nature/water-cycle/ }

Dana Buskovitz 2023, Study.com academy lessons, accessed 6-16-2024

{https://study.com/academy/lesson/the-water-cycle-precipitation-condensation-and-evaporation.html}

A printer friendly glossary for you.

The Water Cycle: Sometimes called the **Hydrologic Cycle**, it is the continuous movement of water through and around the earth.

Water Vapor: The gas formed from heated or boiled water.

Evaporation: The sun heats liquid water and it turns into water vapor.

Condensation: Water vapor cools and forms liquid water.

Precipitation: Water returns to the earth's surface as rain, sleet, snow, or hail.

Transpiration: The sun pulls water from the plants through leaf openings called stomata.

Runoff: Water, pulled by gravity, gets pulled over the ground.

Infiltration: When water seeps into the ground.

Storage: Water is stored three basic locations: in the atmosphere; on the surface of the earth as oceans, lakes, and rivers; and in the ground.

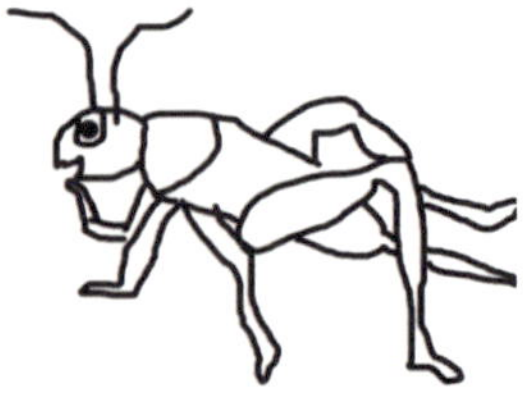

Acknowledgements

To Rachel and Ryan, and the many hours spent learning science at Deer Park Elementary school;

To my husband Dale, for your patience and editorial skills;

To my network of author friends who gave honest opinions:

Thank you!

This book is complete because of your help.

Sonja McGiboney, a native of Pennsylvania now living in Alabama, has authored over twenty-five books for children.

In 2015, Sonja, and her husband Dale, adopted a puppy and named her Jazzy. It was only natural to document Jazzy's antics. After a year of pictures, she created *Growing Up Jazzy*.

Jazzy is now the star of twenty-three concept books for children that combine beautiful photography with lyrical writing and rhyme.

The Water Cycle with Olly and Carl, is the second book written by Sonja that does not star Jazzy, but it is the first book to use her drawings instead of photographs.

Sonja loves photography, writing, spending time with family, and, like Carl, dancing in the rain.

Things you can do!

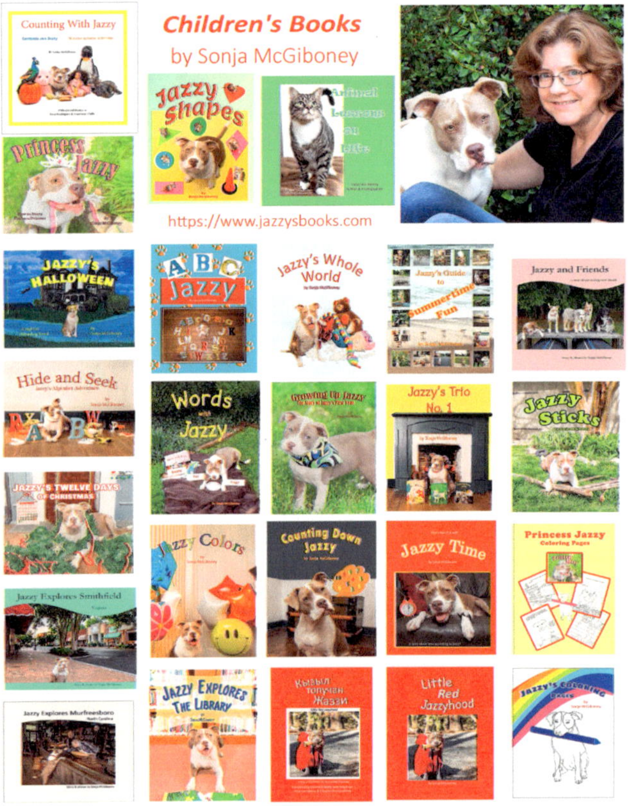

Sign up:

Subscribe to the montly newsletter and get a free coloring page each month. Know the latest contests, stories, and new releases.

https://www.jazzysbooks.com/

Write a review:

If you like this book, and Sonja's other books, let the world know.

Give a teacher a book:

Instead of candy or flowers, that won't last, buy a book to give to your child's teacher for their classroom. That way, more children can enjoy reading.